JAN 2022

TOOLS FOR CAREGIVERS

- **F&P LEVEL:** B
- **WORD COUNT:** 25
- **CURRICULUM CONNECTIONS:** patterns, nature

Skills to Teach

- **HIGH-FREQUENCY WORDS:** I, more, see
- **CONTENT WORDS:** diamonds, lines, ripples, sand, spots, tracks
- **PUNCTUATION:** periods
- **WORD STUDY:** /k/, spelled ck (tracks); long /e/, spelled ee (see); multisyllable word (diamonds)
- **TEXT TYPE:** factual description

Before Reading Activities

- Read the title and give a simple statement of the main idea.
- Have students "walk" through the book and talk about what they see in the pictures.
- Introduce new vocabulary by having students predict the first letter and locate the word in the text.
- Discuss any unfamiliar concepts that are in the text.

After Reading Activities

Explain that tracks are the prints animals leave behind as they walk or move on the ground. Ask readers to make their own tracks. If there is not sand, dirt, mud, or snow outside to make footprints, show them more images of footprints or tracks. Do readers notice the pattern?

Tadpole Books are published by Jump!, 5357 Penn Avenue South, Minneapolis, MN 55419, www.jumplibrary.com

Copyright ©2021 Jump. International copyright reserved in all countries. No part of this book may be reproduced in any form without written permission from the publisher.

Editor: Jenna Gleisner **Designer:** Michelle Sonnek

Photo Credits: arka38/Shutterstock, cover, 2bl, 12–13; texcroc/iStock, 1; San Hoyano/Shutterstock, 2mr, 3; GybasDigiPhoto/Shutterstock, 2ml, 4–5; INTERTOURIST/Shutterstock, 2br, 6–7; Eric Isselee/Shutterstock, 2tl, 8–9; Chantelle Bosch/Shutterstock, 10–11; Ovidiu Hrubaru/Shutterstock, 2tr, 14–15; Kokhanchikov/Shutterstock, 16.

Library of Congress Cataloging-in-Publication Data
Names: Nilsen, Genevieve, author.
Title: Patterns in the desert / by Genevieve Nilsen.
Description: Minneapolis: Jump!, Inc., 2021. | Series: Patterns in nature | Includes index. | Audience: Ages 3–6
Identifiers: LCCN 2020023881 (print) | LCCN 2020023882 (ebook) | ISBN 9781645277569 (hardcover)
ISBN 9781645277576 (paperback) | ISBN 9781645277583 (ebook)
Subjects: LCSH: Deserts—Juvenile literature. | Pattern perception—Juvenile literature.
Pattern formation (Physical science)—Juvenile literature.
Classification: LCC QH88 .N55 2021 (print) | LCC QH88 (ebook) | DDC 577.54—dc23
LC record available at https://lccn.loc.gov/2020023881
LC ebook record available at https://lccn.loc.gov/2020023882

PATTERNS IN NATURE

PATTERNS IN THE DESERT

by Genevieve Nilsen

TABLE OF CONTENTS

Words to Know 2

In the Desert 3

Let's Review! 16

Index 16

WORDS TO KNOW

diamonds

lines

ripples

sand

spots

tracks

IN THE DESERT

I see sand.

I see ripples.

I see tracks.

snake

I see diamonds.

I see more tracks.

cactus

I see spots.

I see lines.

15

LET'S REVIEW!

A pattern is a design that repeats. What pattern do you see here?

INDEX

diamonds 9
lines 15
ripples 5

sand 3
spots 13
tracks 7, 11